THE COOLEST CHINESE Inventions

Joseph Kampff

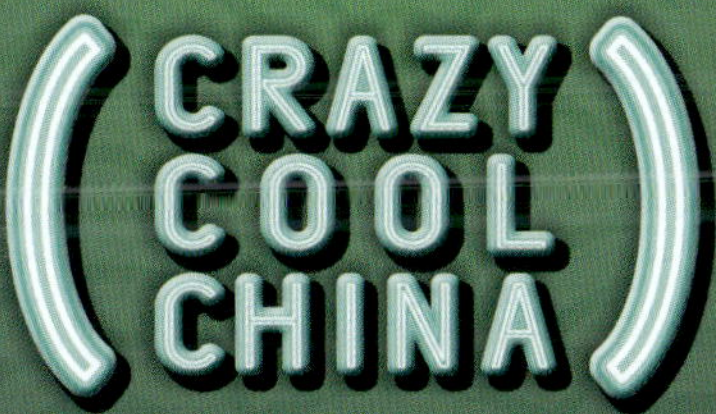

NEW YORK

SinolinguA
华语教学出版社

BEIJING

Published in 2022 by The Rosen Publishing Group, Inc.
29 East 21st Street, New York, NY 10010

Jointly published in 2022 by Sinolingua Co.,Ltd., Beijing, China, and The Rosen Publishing Group, Inc., New York, New York, United States.

First Edition

Editor: Joseph Kampff
Designer: Rachel Rising

Photo credits: Cover, Kach233/Shutterstock.com; cover, pp. 1, 3–48 Sylfida/Shutterstock.com; cover, p. 1 CkyBe/Shutterstock.com; pp. 4, 6, 10, 14, 18, 22, 26, 30, 34, 38, 42 ViewStock/Shutterstock.com; p. 5 lapas77/Shutterstock.com; p. 7 Andrea Paggiaro/Shutterstock.com; p. 8 © Look and Learn/Bridgeman Images; p. 11 Pictures from History/CPA Media Pte Ltd/Alamy Stock Photo; p. 12 Bikeworldtravel/Shutterstock.com; p. 15 Luca Lorenzelli/Shutterstock.com; p. 16 https://commons.wikimedia.org/wiki/File:Beijing.China_printing_museum.Plate_of_Paper_money.Northern_Song_Dynasty.jpg; p. 19 https://commons.wikimedia.org/wiki/File:Pen_ts%27ao,_woodblock_book_1249-ce.png; p. 20 https://commons.wikimedia.org/wiki/File:Jingangjing.jpg; p. 23 Stephen Dalton/Avalon.red/Alamy Stock Photo; p. 24 ZengTengbo/Shutterstock.com; p. 27 https://commons.wikimedia.org/wiki/File:Zhou_Wenju_%E9%87%8D%E5%B1%8F%E4%BC%9A%E6%A3%8B%E5%9B%BE_Palace_Museum,_Detail_of_Go_Players.jpg; p. 28 Ahn Young-joon/Associated Press/AP Images; p. 31 kai keisuke/Shutterstock.com; p. 32 margaret tong/Shutterstock.com; p. 35 Zhang chi sd/Zhang chi sd - Imaginechina/AP Images; p. 36 https://commons.wikimedia.org/wiki/File:Luban_sculpture_weifang_2010_06_06.jpg; p. 39 gyn9037/Shutterstock.com; p. 40 Lian guoqing/Lian guoqing - Imaginechina/AP Images; p. 43 GuoZhongHua/Shutterstock.com.

Some of the images in this book illustrate individuals who are models. The depictions do not imply actual situations or events.

Library of Congress Cataloging-in-Publication Data

Names: Kampff, Joseph.
Title: The coolest Chinese inventions / Joseph Kampff.
Description: New York : Rosen Young Adult, 2022. | Series: Crazy cool China | Includes glossary and index.
Identifiers: ISBN 9781499472363 (pbk.) | ISBN 9781499472370 (library bound) | ISBN 9781499472387 (ebook)
Subjects: LCSH: Technology–China–History–Juvenile literature. | Science–China–History–Juvenile literature. | Inventions–China–History–Juvenile literature.
Classification: LCC T27.C5 K36 2022 | DDC 609.31–dc23

Manufactured in the United States of America

CPSIA Compliance Information: Batch #CSPK23. For further information, contact Rosen Publishing, New York, New York, at 1-800-237-9932.

Contents

A World Leader in Innovation

Chinese civilization had existed for a long time when its written history began around 3,500 years ago. This makes China one of the oldest continuous civilizations in the world. Unlike the United States, for example, which is not even 250 years old and is made up of people and cultures from all around the world, the China of today is built on the foundation of ancient Chinese culture. While many Westerners look to ancient Greece for the roots of their culture, they owe quite a lot to China.

The compass, gunpowder, papermaking, and printing: these are the Four Great Inventions celebrated in China. Europeans originally believed that three of these inventions—the compass, gunpowder, and printing—came from Europe. But in the 16th century, Europeans learned that the Chinese were using these technologies long before. The British **sinologist** (a person who studies Chinese culture) Joseph Needham helped spread the idea of the Four Great Inventions in the 20th century. The Four Great Inventions are so important to Chinese identity that the Chinese used them as a major theme for the 2008 Olympics in Beijing.

This is particularly true in the realm of technology. Many of the world's most important and well-known inventions came from China. These range from the Four Great Inventions—the compass, gunpowder, papermaking, and woodblock printing—to one of the world's most expensive and luxurious **textiles**, silk; the most complex board game, Go; and everyday objects such as kites and umbrellas. China has been and continues to be a world leader in technological innovation.

The Terracotta Army is from the tomb of China's first emperor, Qin Shi Huang, who reigned from 221 to 210 BCE.

The Compass

A compass is a device that shows the **cardinal directions** for **navigation** or determining one's geographical position. Magnetic compasses are the most common compasses. They have a magnetized needle that responds to Earth's magnetic field to pull one end of the needle in the direction of Earth's magnetic north and the other end toward the magnetic south. Although there have been many improvements to the magnetic compass since it

The earliest compasses were made in China out of a magnetic mineral called magnetite. When shaped into a needle and suspended in the air or floating in a bowl of water, magnetite lines up with Earth's magnetic field and points north and south. The Chinese shaped magnetite into spoons, turtles, and fish that pointed south when floating in a bowl of water. These south-pointing fish were used to align houses according to the principles of **feng shui**, which would allow positive energy to flow into them.

was invented, the basic principle that makes it work remains the same today.

The earliest magnetic compasses were invented in China during the Han dynasty (206 BCE–220 CE). These compasses were used for **divination**—using **supernatural** means to gain knowledge of the future—and for feng shui, a system for harmonizing environments with the spiritual forces within them. These early compasses are known as *luo pan*, which refers to the union of the heavens and Earth.

The South-Pointing Fish is an early Chinese magnetic compass that was used to align buildings according to feng shui.

The mariner's compass was a crucial invention for navigating the seas.

Before radar, satellites, or even reliable maps, ship captains determined their course by looking at landmarks and the sun and stars. This wasn't reliable far out at sea or when it was cloudy. The invention of the mariner's compass changed navigation forever by allowing people to travel the seas with more certainty than ever before. In 1117 CE, the Chinese historian Zhu Yu was the first to record the use of a compass for navigation: "The navigator knows the geography, he watches the stars at night, watches the sun at day; when it is dark and cloudy, he watches the compass."

The Chinese were also the first to use the magnetic compass for navigation. They were probably using the compass for this purpose by the 11th century CE. Western Europeans, on the other hand, didn't start using the compass for navigation until the 12th century.

Before the invention of the compass for navigation, travelers had to get their **bearings** using landmarks, the sun, and the stars. Before the invention of the compass, most ships stayed close to land so they wouldn't get lost. The Chinese use of the compass for navigation dramatically changed the way people traveled over the seas.

Gunpowder

Few inventions have had such a dramatic effect on world history as gunpowder. After its discovery by Daoist priests in the 9th century CE, at the latest, gunpowder changed warfare forever. The irony is that the Daoist priests who discovered gunpowder weren't thinking about weapons at all. Rather, they were looking for an **elixir** that would extend life.They found the opposite.

The invention of gunpowder in China soon led to the invention of fireworks. The original fireworks were unlike what you might see during a New Year's celebration or other holiday. They weren't colorful and they weren't shot into the air. The first fireworks were simply gunpowder packed into bamboo or paper tubes that were thrown into a fire. The noise they made was believed to scare away evil spirits.

Gunpowder is a highly flammable mixture of saltpeter (or potassium nitrate), sulfur, and charcoal. It's not surprising that saltpeter would be used to create an elixir. Saltpeter had been used by the

Chinese for medicinal purposes since ancient times. And the Chinese word for gunpowder, *huoyao*, means literally "fire medicine."

There's a common misconception that gunpowder was at first only used by the Chinese for entertainment. But the military uses of gunpowder were obvious to the Chinese from the beginning. There's evidence of gunpowder **incendiaries** being used by Chinese warriors against their enemies as early as 900 CE. China's armies were the first in the world to have firearms and gunpowder.

The Chinese fire lance is the ancestor of modern firearms.

Fireworks are an important element of celebrations around the world, including Chinese New Year and the Fourth of July in the United States.

Around 1200 CE, the Chinese used gunpowder to invent the first rockets. Chinese armies attached tubes of gunpowder to the arrows they shot at their enemies. But they soon realized that the gunpowder could launch the tubes by itself. This is how the first rockets were invented. While rockets were originally meant to fire **projectiles** at enemies during battle, they were also used to launch fireworks into the sky.

The Chinese eventually used gunpowder in the development of other firearms, including guns and cannons. In fact, the Chinese didn't just invent gunpowder, they also invented the first gun: the Chinese fire lance. The fire lance was a long spear that had a firework attached to the end of it that shot a small projectile or poison when lit. Because the projectile had a short range, the fire lance was used for close combat.

The use of gunpowder has spread all around the world. And although we tend to think of gunpowder for its military uses, it has been a fundamental component in agricultural and land development, the creation of railroads and tunnels, and the construction of power-generating dams.

Papermaking

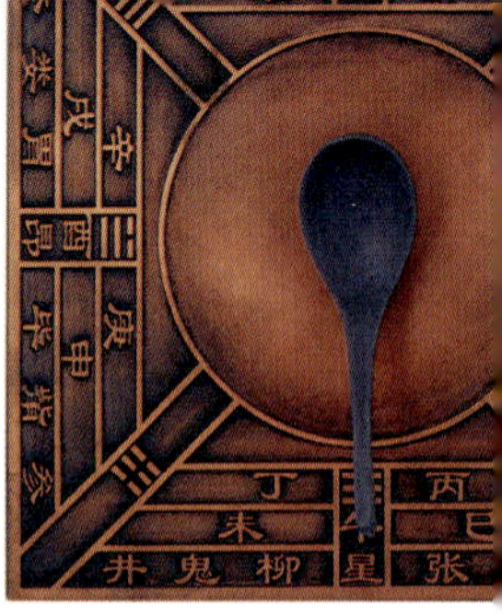

The traditional story about the invention of paper is that around 105 CE, Cai Lun created it out of tree bark, **hemp**, old rags, and fishing nets. But **archaeological** evidence shows that the Chinese were using paper in the 2nd century BCE. Before the invention of paper, writing in China was done on bamboo, strips of wood, or silk. The problems with these were that bamboo and wood were heavy and bulky, and silk was very expensive to produce. Paper was relatively inexpensive, light, and easy to carry.

Although the Silk Road is mostly known for the trade in silk, paper and, eventually, the process of making paper are some of the most important things carried along the Silk Road. The spread of paper and papermaking technology improved our ability to record and transmit information throughout the world. Even after the process of papermaking was known outside of China, Chinese paper was still a valuable **commodity** because it was the highest quality.

The earliest paper was made from hemp. Clothing at that time was made from hemp, and it's possible that paper was invented when hemp clothing was left too long in water after washing. This left a residue in the water that could be compressed into paper. Cai Lun experimented with different plant fibers and other materials to refine the papermaking process. The goal

Although the technique for making paper by hand is centuries old, it's still practiced today.

The world's first paper money, called *jiaozi*, was first used during the Song dynasty (960–1279 CE).

The Chinese were the first people to use paper money. The idea came from merchants and government officials who didn't want to carry large quantities of coins and other forms of **currency**. Before paper money, they used receipts called "flying money"—because it could get caught in the wind and fly away—which was redeemed for coins later. Paper money became a common form of currency in China during the Song dynasty (960–1279 CE).

was to make the highest quality paper with the least expensive material. For centuries, paper was made from rattan, a kind of palm. But the demand for paper was high, so slow-growing rattan was replaced by bamboo before it was replaced by the bark of the mulberry tree. The process of paper production began to spread to Europe in the 8th century.

The invention of paper was central to the spread of literacy and literature. And with the Chinese invention of woodblock printing, the value of paper for writing and making books was greater than ever. But paper had many more uses for the Chinese than writing. **Cartographers** were able to make high-quality portable maps for military purposes. Paper was used to package tea, to make screens, and to make money.

Woodblock Printing

With the invention of paper, printing wasn't far behind. The Chinese had invented woodblock printing by the end of the 2nd century CE. At its most basic, printing requires paper, ink, and a surface with text carved in relief. Relief carving is when images or letters are carved out of a material (stone, wood, etc.) while leaving the material in place as a solid background. An early form of printing involved Buddhist texts inscribed on stone pillars. People would put ink on the text and press paper to them to print copies.

Although it's impossible to say exactly when ink was invented, we do know that it was invented thousands of years ago by the Chinese and the Egyptians. Chinese ink (also known as India ink) is a special kind of ink that is in a solid form called inkstick. Before using it for drawing, writing, or printing, Chinese ink is ground down and mixed with water. Today, it's used mostly for **calligraphy** and other art projects.

Carving texts in stone takes a lot of time and effort. Therefore, in the 6th or 7th century, the Chinese started carving texts into wood blocks for printing. Using wooden blocks made printing much easier and faster. To prepare a wood block for printing, a scribe would write the text on a piece of paper with ink. Before the

Although still time consuming, woodblock printing made the creation of printed texts much more efficient.

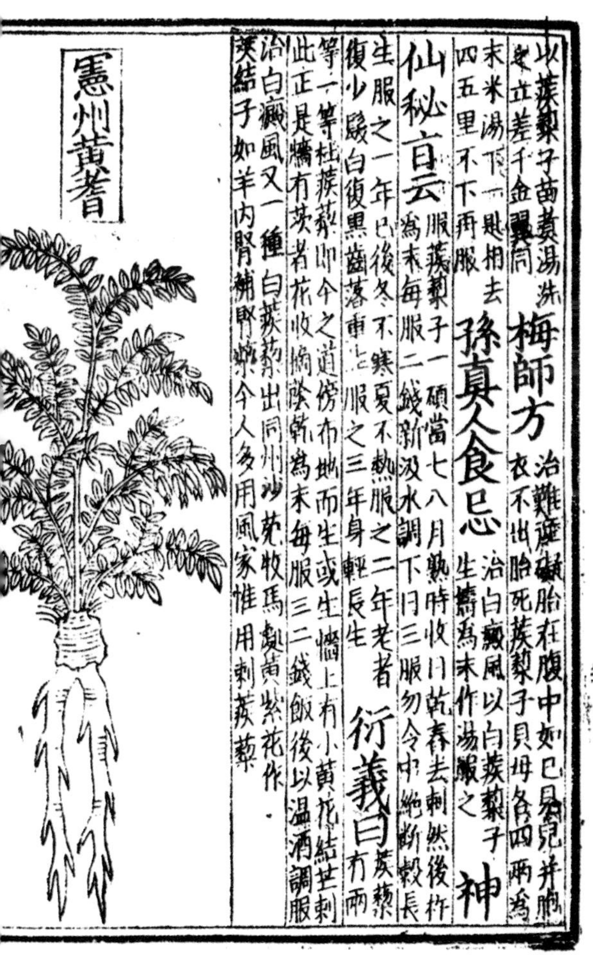

The Diamond Sutra, dated 868 CE, is the world's oldest printed book.

While woodblock printing was a major improvement over carving texts in stone, it was time consuming. To make the printing process faster, movable-type printing was invented in China by an expert woodblock printer named Pi Sheng. Sometime between 1041 and 1048, Pi Sheng came up with the idea of using clay to make blocks of type for Chinese characters. These were stuck to an iron plate that was used to make the print. Once the print was made, the clay blocks could be removed and replaced with other blocks.

ink dried, he would press it against the wooden block. Then the parts of the wood around the text would be carved out, leaving the text in relief so ink could be applied to the block before pressing it to paper for printing. If you've ever seen an old-fashioned rubber stamp, you can imagine how this works.

The advent of woodblock printing changed the way information was recorded and circulated. Woodblock printing spurred on the paper production industry in China and soon became essential for business operations and transactions. Early woodblock printing was mostly used for medical and agricultural texts. And in 868 CE, the world's first book, *The Diamond Sutra*, was made in China using woodblock printing.

Silk

China not only invented silk production (also known as **sericulture**), but it also kept the process a secret for thousands of years. The earliest archaeological evidence of silk in China goes back around 8,500 years. Other parts of the world didn't learn about sericulture until knowledge of the process traveled along the Silk Road to India in the 2nd century CE. Before then, all the silk in the world came from China.

There's a legend that sericulture was invented by Empress Leizu in the 27th century BCE. Empress Leizu was having a cup of tea under a mulberry tree in the royal garden when a moth's cocoon fell into her cup. She took it out and saw a strand of fiber coming from the cocoon. She pulled on this strand until it filled the entire royal garden. When she wove the strand into fabric, silk was invented.

Silk is a fine textile, or fabric, that's woven from the fiber strands of silkworm **larvae**, which are caterpillars of silk moths. The cocoon of a single silkworm has a thousand yards of thread that can be used to make silk fabric. This may seem like a lot, but silk is a delicate fiber. It takes about 2,500 silkworm larvae to produce a pound of silk.

Silk is an entirely natural fabric. It all begins with the cocoon of the silk moth.

Silk is a customary fabric for *hanfu*, the traditional clothing of the Han people in China.

Established during the Han dynasty (206 BCE–220 CE), the Silk Road created essential trade routes connecting China and the West. It was called the Silk Road, or Silk Routes, because silk was one of the most important goods transported along the routes. Other goods, such as tea, spice, paper, and gunpowder, were traded along the Silk Road, but the exchange of culture was one of the most significant effects of the Silk Road.

Silk fabric can be used for clothing such as fine shirts and wedding dresses, accessories like ties and scarves, **tapestries**, bedding, and surgical **sutures**. If you've ever had a cut that needed stitches, there's a good chance they were silk.

Silk is a luxury fabric. It's closely associated with royalty and has been used as a form of currency. Because it was highly valued for trade, silk gave its name to the Silk Road, the network of routes across central Asia that traders used to travel between China and other parts of the world.

Silk production is still going strong in China today. It produces around 78 percent of the world's silk.

The Game of Go

Go might be the world's oldest board game. Although the exact date isn't known, Go is likely to have been invented in China about 4,000 years ago. Mastering the game of Go was among the "Four Accomplishments" that a Chinese gentleman or lady must achieve (along with playing a stringed instrument, calligraphy, and painting) in ancient China. Go is one of the most complex games in the world, but its rules are so simple anyone can play.

There are several legends about the creation of Go. One of the most prominent is that it was invented by the legendary emperor Yao, who is said to have lived from around 2356 to 2255 BCE. The story goes that he designed the game to teach his son discipline, concentration, and balance. Another popular story is that it was created by military strategists to plan attacks.

In China, Go is called Wei Qi, which means "surrounding game." This name aptly expresses the object of Go. A normal-sized Go board has 19 by 19 lines (although faster games can be played on smaller boards). The game starts with an empty board. Two players take turns placing pieces on the intersecting

Mastery of the game of Go was one of the signs of a gentleman in ancient Chinese culture.

Today, artificial intelligence can beat the world's best Go players.

Although Go is often compared to chess, Go is much more complex. In 1997, a chess computer called Deep Blue was the first **artificial intelligence** computer to beat the world's reigning champion chess player. Deep Blue could calculate up to 200 million chess positions per second. Go, on the other hand, might have more possible board configurations than there are atoms in the known universe. It wasn't until 2015 that an AI program called AlphaGo beat a human Go champion for the first time.

lines. These pieces, called stones, are either black or white. The player with the black stones goes first. The goal is to use your stones to surround space on the board so it cannot be occupied by the other player's stones. This is your territory. You can also take one of the other player's stones by surrounding it with your pieces. Points are counted at the end of the game, with one point for each vacant space within your territory and one point for each prisoner you've taken. At the end of the game, the player with the most points wins.

Go has traveled far from China and is immensely popular in Japan and Korea, and it's played all around the world.

Tea Production

It's not too much of a stretch to say that just about everybody likes tea. It's one of the most popular beverages in the world. And the Chinese have been **cultivating** and drinking tea longer than anyone else.

Tea drinking in China goes back around 4,000 years. There are several myths about the origin of tea drinking. One is that Emperor Shennong, who ruled around 2800 BCE, set up camp with his entourage under a tree and set a pot of

When you think about kung fu, Bruce Lee, Jackie Chan, and Jet Li movies probably come to mind. But in China, kung fu is a term that describes anything that takes time and practice to master, and it applies as much to tea as to martial arts. During a kung fu tea ceremony, rare and expensive leaves are steeped in small pots to maximize the taste and aroma of the tea. Kung fu tea means making tea with skill.

water to boil so it would be purified for drinking. The heat from the fire dried the leaves of the tree, and some of them fell into the water, turning its color to gold. Smelling the wonderful aroma of the water, the emperor tasted it and exclaimed, "T'sa"—God-like—at its flavor and energizing effect. The word *cha* is still used for tea in China today.

Longjing is a hand-produced green tea that is very popular in China.

The kung fu tea ceremony takes drinking tea to the level of fine art.

Tea houses are as important to Chinese culture as tea itself. People go to tea houses to relax, meet friends, have conversations, do business, and, of course, drink tea. Chinese tea houses are a lot like coffee shops. But China's tea houses have been around much longer than coffee shops. The first coffee shop was established in Turkey in the 15th century. Tea houses started to appear in China during the Tang dynasty (618–907 CE) and spread widely during the Song dynasty (960–1279 CE). They've been popular ever since.

Archaeological evidence shows that tea was used primarily by the wealthy as medicine during the Han dynasty (206 BCE–220 CE), but it wasn't until the Tang dynasty (618–907 CE) that tea culture in China as we know it really came alive.

Chinese monks were some of the first people to drink tea regularly. They used it to stay awake while meditating for a long time. In fact, most of what we know about early tea culture in China comes from Lu Yu, an orphan who was raised in a Buddhist monastery. He wrote a book called *The Classic of Tea*, which explains how tea was grown, prepared, and consumed during the Tang dynasty.

Since its beginnings in China, tea culture has traveled all over the world. In fact, over two-thirds of the world's population enjoys tea. It's the second most consumed beverage, behind water.

Kites

The kite was the world's first aircraft. Kites were invented in China over 2,000 years ago. The earliest kites were made of wood and cloth and shaped to resemble birds. They were used by armies to measure distances and to check the speed and direction of the wind. Later kites were made of paper and used for communication. During the Tang dynasty (618–907 CE), kite construction became more refined, with kite makers using silk, paper, and bamboo, and flying kites became a hobby.

Although kites usually bring to mind a breezy day in the park or at the beach, kites were originally used primarily for warfare. The earliest kite was invented in China by Lu Ban during a time known as the Spring and Autumn Period (770–476 BCE). These kites resembled birds in the sky and were used to measure the distances armies had to travel. Later, the incendiary "divine flying fire crow" was flown over enemies and ignited to cause fires in enemy camps.

In 1282, the Italian explorer Marco Polo saw a man strapped to a large kite flying in the sky over Weihai. The goal was to determine if the winds were favorable for sailing. When Marco Polo returned to Europe, he brought a Chinese kite with him. This is how the art of kite making spread from China to the world.

At the 2021 International Weifang Kite Festival, the world's largest kite was flown.

Lu Ban is honored in Weifang as the inventor of the "bamboo sparrow," an early kite.

Lu Ban was an engineer, carpenter, and inventor who lived in China during the Zhou dynasty (1046–256 BCE). He's known for inventing the "cloud ladder," which is a folding ladder that was used for scaling walls in **siege warfare**, grappling hooks and rams used in naval warfare, and the first kite: a bird made of bamboo—a "bamboo sparrow"—that could stay in the air for three days. In Chinese folk religion, Lu Ban is the god of carpentry.

Kites are much more than a children's toy in China. They're an art form. Chinese kites are often large, colorful, and elaborate. They depict creatures from Chinese mythology and legend. The most popular and recognizable Chinese kite today is the centipede kite. Centipede kites are large, three-dimensional kites with dragon heads and centipede bodies.

Kites are such an important part of Chinese culture that the city of Weifang is the location of the Weifang World Kite Museum and holds the annual International Weifang Kite Festival. In 2021, the world's largest kite flew at the festival. The kite, which had the head of a dragon and the body of a centipede, was over 918 feet (280 meters) long.

Umbrellas

Human beings have used umbrellas to keep the sun and rain off their heads for a *long* time. The earliest umbrellas were invented in China around 4,000 years ago. There's evidence of umbrellas being used in Egypt, Assyria, and Greece around the same time. The first umbrellas were used to protect people from the sun. It wasn't until the Chinese invented the first waterproof umbrellas that they could be used in the rain.

Chinese folklore gives two stories of the invention of the umbrella. The first tells of a boy walking through a village in the rain. To keep the rain off, he held a large lotus leaf over his head. This inspired the creation of the umbrella. The second story tells of an army traveling in the heat. The soldiers used lotus leaves to protect their heads from the sun. This inspired the creation of the parasol. In general, umbrellas protect from rain and parasols protect from the sun.

The most recognizable Chinese umbrellas are made of paper or silk. Silk is the most expensive material for umbrella making, and silk umbrellas are considered the most desirable. But silk is difficult to work with and to maintain.

The materials for paper umbrellas are less costly and can be waterproofed using **tung oil**. But that doesn't mean making paper umbrellas is simple. There are 80 processes a skilled craftsman goes through to make a paper umbrella. The paper used for high-quality Chinese umbrellas is thin but strong.

At the head of the Terracotta Army is a driver with an umbrella that was used for shading him from the sun.

The traditional methods for producing oil-paper umbrellas are still practiced in China today.

In Chinese culture, umbrellas are not only used to protect people from the rain and sun. They also have **aesthetic** value. Oil-paper umbrellas have been produced in China since the Eastern Han dynasty (25–220 CE) by painting the paper with oil made from the fruit of the tung tree. This oil makes the paper waterproof. Artists paint traditional images, such as birds, flowers, and landscapes, on the umbrellas. Oil-paper umbrellas are an important element of traditional Chinese weddings because they are believed to protect against bad luck.

Once the paper has been waterproofed, it's ready to be decorated. Chinese oil-painted umbrellas are works of art. The subjects painted on the umbrellas include images from the natural world, such as birds, flowers, and landscapes, as well as scenes from Chinese literature.

During the Tang dynasty (618–907 CE), China began to have more contact with other Asian countries. As a result, oil-paper umbrellas started to be produced in Japan, Korea, Vietnam, Thailand, and Laos.

Although oil-paper umbrellas are used less in the rain now than in the past, they continue to be important cultural and symbolic objects today. Red oil-paper umbrellas are often given as gifts for weddings, birthdays, and other significant events.

The Four New Inventions

Most of the inventions in this book have been around for a long time. But China is a very modern nation that focuses on technological advancement. The Four New Inventions refer to the Four Great Inventions while emphasizing modern technologies. The Four New Inventions are high-speed rail, mobile payment, e-commerce, and bike sharing.

China has the largest network of high-speed railways. High-speed trains, which top out at 217 mph (350 kph) connect all of China's major cities. And all of China's 23,500 miles (37,900 kilometers) of high-speed railways have been built since 2008.

When it comes to paying for goods and services, China went straight from cash to mobile payment. People in China use the apps Alipay and WeChat to pay for almost everything. Around 90% of Chinese people use mobile payment daily.

E-commerce—buying and selling things online—is huge in China. Over half of the world's e-commerce transactions come from China. China's massive share in e-commerce sales comes mostly from Tmall.com, which is the most-visited website in China and the third most visited in the world.

China's a world leader in mobile payment. It's no surprise, then, that "dockless" bike share programs are popular in China. Dockless bikes are unlocked using a mobile app. The bikes can be rented and dropped off anywhere. This is a great solution for commuters who take the bus or subway to work or school and need a ride to and from the station. Today, e-bike share programs are overtaking conventional bikes in China.

China has more high-speed railways than anywhere else in the world.

Glossary

aesthetic Having to do with art and beauty.
archaeological Having to do with the study of human history by excavating, or digging, sites of human civilization.
artificial intelligence A computer's ability to do thing that are usually associated with human intelligence.
bearings A person's understanding of their geographical position.
calligraphy Elaborate and artistic handwriting.
cardinal direction One of the four main points on a compass: north, south, east, and west.
cartographer A person who makes maps.
commodity A good that can be bought and sold.
cultivate To prepare and use soil for growing plants.
currency A system of money or form of payment.
divination Predicting the future.
elixir A medicinal or magical potion.
feng shui The practice of organizing spaces in accordance with spiritual forces.
hemp An Asian herb that is grown for its fiber.
incendiary A weapon intended to start a fire.
larva The wormlike form of an insect when it hatches from an egg.
navigation The science of determining the course of a ship or other type of craft.
projectile A weapon that is self-propelled.
sericulture The method of producing silk from silkworms.
siege warfare A strategy in warfare that involves surrounding a town for a long period of time.
sinologist A person who studies the culture of China.
supernatural Relating to forces or beings beyond what can be known scientifically.
suture A strand that's used to sew parts of the body; a stitch.
tapestry A woven textile that's hung for aesthetic enjoyment.
textile Woven or knit cloth.
tung oil The oil from the seeds of the tung tree, which is used for waterproofing.

Bibliography

Alexa. "Tmall.com Competitive Analysis, Marketing Mix and Traffic." Accessed October 4, 2021. https://www.alexa.com/siteinfo/tmall.com.

Asian Art Museum. "The Invention of Woodblock Printing in the Tang (618–906) and Song (960–1279) Dynasties." Accessed October 4, 2021. https://education.asianart.org/resources/the-invention-of-woodblock-printing-in-the-tang-and-song-dynasties/.

Beijing Tourism. "Things You May Not Know About Chinese Oil-Paper Umbrellas." August 9, 2017. http://english.visitbeijing.com.cn/a1/a-XCX1LD386656CF2828B1A6.

British Go Association. "How to Play." Accessed October 4, 2021. https://www.britgo.org/intro/intro2.html.

Britt, Kenneth W. "Papermaking." Encyclopedia Britannica. January 10, 2020. https://www.britannica.com/technology/papermaking.

Cartwright, Mark. "Paper in Ancient China." World History Encyclopedia. Accessed October 4, 2021. https://www.worldhistory.org/article/1120/paper-in-ancient-china/.

ChinaCulture.org. "Who Invented Kite?" December 30, 2014. http://en.chinaculture.org/2014-12/30/content_589624.htm.

China Highlights. "Chinese Kites—History and Culture." October 2, 2021. https://www.chinahighlights.com/travelguide/culture/kites.htm.

China Today. "Oil-Paper Umbrella: Artwork and Accessory." May 15, 2018. http://www.chinatoday.com.cn/ctenglish/2018/ich/201805/t20180515_800129484.html.

ChinaTravel.com. "Chinese Kite." Accessed October 4, 2021. https://www.chinatravel.com/culture/chinese-kite.

ChinaTravel.com. "Chinese Paper Umbrella." Accessed October 4, 2021. https://www.chinatravel.com/culture/chinese-paper-umbrella.

ChinaTravel.com. "History of Chinese Silk Production." Accessed October 4, 2021. https://www.chinatravel.com/culture/history-of-chinese-silk-production.

Choi, Charles Q. "Oldest Evidence of Silk Found in 8,500-Year-Old Tombs." Livescience. January 10, 2017. https://www.livescience.com/57437-oldest-evidence-of-silk-found-china.html.

DeepMind. "AlphaGo." Accessed October 4, 2021. https://deepmind.com/research/case-studies/alphago-the-story-so-far.

Dethlefsen & Balk. "Legend of Teas." Accessed October 4, 2021. https://www.dethlefsen-balk.de/ENU/10730/Legende_vom_Tee.html.

Dugdale-Pointon, TDP. "Li Hua Ch'iang (Fire-Lance)." History of War. Accessed October 4, 2021. http://www.historyofwar.org/articles/weapons_firelance.html.

Encyclopedia Britannica. "Go." April 24, 2017. https://www.britannica.com/topic/go-game.

Encyclopedia Britannica. "Silk." March 7, 2019. https://www.britannica.com/topic/silk.

Encyclopedia Britannica. "Zhou Dynasty." May 12, 2020. https://www.britannica.com/topic/Zhou-dynasty.

Fairbairn, John. "The Legends of the Sage Kings and Divination." GOBASE.ORG. Accessed October 4, 2021. https://gobase.org/reading/history/china/?sec=part-2.

Feng Shui Institute. "The Luo Pan Chinese Compass." Accessed October 4, 2021. https://www.feng-shui-institute.org/Feng_Shui/luopan.html.

Global Times. "The Origins and Traditions of Chinese Oil-Paper Umbrellas." August 8, 2017.

Bibliography

https://www.globaltimes.cn/content/1060250.shtml.

Goodrich, Joanna. "How IBM's Deep Blue Beat World Champion Chess Player Garry Kasparov." IEEE Spectrum. January 25, 2021. https://spectrum.ieee.org/the-institute/ieee-history/how-ibms-deep-blue-beat-world-champion-chess-player-garry-kasparov.

Guinness World Records. "First Paper Money." Accessed October 4, 2021. https://www.guinnessworldrecords.com/world-records/first-paper-money.

Hamilton, Mae. "Lu Ban." Mythopedia. Accessed October 4, 2021. https://mythopedia.com/chinese-mythology/gods/lu-ban/.

Harder, Jeff, and Charise Cunningham. "Who Invented the First Gun?" How Stuff Works. August 31, 2021. https://science.howstuffworks.com/innovation/inventions/who-invented-the-first-gun.htm.

HiSoUR.com. "Origin of the Umbrella, China Umbrella Museum." Accessed October 4, 2021. https://www.hisour.com/origin-of-the-umbrella-china-umbrella-museum-48261/.

Hong, Yaobin. "World's Largest Kite Wows Visitors at Weifang International Kite Festival 2021." CGTN. April 19, 2021. https://news.cgtn.com/news/2021-04-19/World-s-largest-kite-takes-to-the-skies-in-E-China-s-Weifang-ZAlmRPoKA0/index.html.

Jakhar, Pratik. "Who Really Came Up with China's 'Four New Inventions'?" BBC. April 3, 2018. https://www.bbc.com/news/world-asia-china-43406560.

Jiang, Fercility. "Chinese Silk—Silk History, Production, and Products." China Highlights. October 2, 2021. https://www.chinahighlights.com/travelguide/culture/chinese-silk.htm.

Jiang, Fercility. "Traditional Chinese Paper Umbrellas: Origins and Making." China Highlights. https://www.chinahighlights.com/travelguide/culture/paper-umbrella.htm.

Jones, Ben. "Past, Present and Future: The Evolution of China's Incredible High-Speed Rail Network." CNN Travel. May 26, 2021. https://www.cnn.com/travel/article/china-high-speed-rail-cmd/index.html.

Joukowsky Institute for Archaeology and the Ancient World. "Gunpowder: Origins in the East." Accessed October 4, 2021. https://www.brown.edu/Departments/Joukowsky_Institute/courses/13things/7687.html.

Karam, P. Andrew. "The Chinese Invent the Magnetic Compass." Accessed October 4, 2021. https://www.encyclopedia.com/science/encyclopedias-almanacs-transcripts-and-maps/chinese-invent-magnetic-compass.

Lechêne, Robert. "The Invention of Printing." Encyclopedia Britannica. October 1, 2020. https://www.britannica.com/topic/printing-publishing/The-invention-of-printing.

Lu, Houyuan, et al. "Earliest Tea as Evidence for One Branch of the Silk Road Across the Tibetan Plateau." National Center for Biotechnology Information, US National Library of Medicine. January 7, 2016. https://www.ncbi.nlm.nih.gov/pmc/articles/PMC4704058/.

Ma, Winston. "Here Are 4 Major Bike-Sharing Trends from China After Lockdown." World Economic Forum. July 22, 2020. https://www.weforum.org/agenda/2020/07/4-big-bike-sharing-trends-from-china-that-could-outlast-covid-19/.

Mark, Joshua J. "Silk Road." World History Encyclopedia. May 1, 2018. https://www.worldhistory.org/Silk_Road/.

Bibliography

Meredith, Anne. "Journey into the World of Chinese Tea." June 30, 2021. https://studycli.org/chinese-culture/tea/.

NASA. "Brief History of Rockets." Accessed October 4, 2021. https://www.grc.nasa.gov/www/k-12/TRC/Rockets/history_of_rockets.html.

New World Encyclopedia. "Compass." Accessed October 4, 2021. https://www.newworldencyclopedia.org/entry/Compass.

New World Encyclopedia. "Go (board game)." Accessed October 4, 2021. https://www.newworldencyclopedia.org/entry/Go_(board_game).

Ovide, Shira. "Don't Even Try Paying Cash in China." *New York Times.* January 28, 2021. https://www.nytimes.com/2020/10/27/technology/alipay-china.html.

Paajanen, Sean. "The Evolution of the Coffee House." The Spruce Eats. February 6, 2019. https://www.thespruceeats.com/evolution-of-the-coffee-house-765825.

People's Daily Online. "Over 70 Percent of Chinese Use Mobile Payments Every Day in 2020." January 14, 2021. http://en.people.cn/n3/2021/0114/c90000-9809611.html.

Rodgers, Alison Ince. "Compass." *National Geographic.* December 3, 2013. https://www.nationalgeographic.org/encyclopedia/compass/.

Rosen, Diana. "The History of Chinese Teahouses." TeaMuse. August 27, 2020. https://www.teamuse.com/article_200828.html.

Salat, Harris. "From a Long Tradition, Small Sips Quickly Brewed." *New York Times.* September 13, 2006. https://www.nytimes.com/2006/09/13/dining/13kungfu.html?.

Stempien, Alexis. "The Evolution of Fireworks." Smithsonian Science Education Center. Accessed October 4, 2021. https://ssec.si.edu/stemvisions-blog/evolution-fireworks.

UNESCO. "Did You Know? The Importance of Paper Making Technology in Cultural Exchange Along the Silk Roads." Accessed October 4, 2021. https://en.unesco.org/silkroad/content/did-you-know-importance-paper-making-technology-cultural-exchange-along-silk-roads.

Wollensak, Daniella. "The Story of Silk." Shen Yen Performing Arts. August 16, 2018. https://www.shenyunperformingarts.org/blog/view/article/e/NOxQ55iXVE8/story-of-silk.

Index